FOREST CREATURES

OPOSSUMS IN THE FOREST

SOPHIE WASHBURNE

Gareth Stevens
PUBLISHING

Please visit our website, www.garethstevens.com. For a free color catalog of all our high-quality books, call toll free 1-800-542-2595 or fax 1-877-542-2596.

Library of Congress Cataloging-in-Publication Data
Names: Washburne, Sophie, author.
Title: Opossums in the forest / by Sophie Washburne.
Description: New York, NY : Gareth Stevens, [2023] | Series: Forest creatures | Includes bibliographical references and index.
Identifiers: LCCN 2022004298 | ISBN 9781538279175 (library binding) | ISBN 9781538279151 (paperback) | ISBN 9781538279168 (set) | ISBN 9781538279182 (ebook)
Subjects: LCSH: Opossums–Juvenile literature.
Classification: LCC QL737.M34 W37 2023 | DDC 599.2/76–dc23/eng/20220315
LC record available at https://lccn.loc.gov/2022004298

Published in 2023 by
Gareth Stevens Publishing
29 East 21st Street
New York, NY 10010

Portions of this work were originally authored by Ada Quinlivan and published as *Opossums*. All new material in this edition was authored by Sophie Washburne.

Editor: Jennifer Lombardo
Designer: Rachel Rising

Photo credits: cover David Louis Econopouly/Shutterstock.com; cover, p. 1 furtseff/Shutterstock.com; cover, pp. 1, 3, 4, 6, 8, 10, 12, 14, 16, 18, 20, 22–24 lavendertime/Shutterstock.com; cover, pp. 1, 3, 6, 8, 10, 12, 14, 20, 22–24 Zvezdesign/Shutterstock.com; p. 4 IrinaK/Shutterstock.com; p. 5 miwa-in-oz/Shutterstock.com; p. 7 BigMouse/Shutterstock.com; pp. 9, 21 Jay Ondreicka/Shutterstock.com; p. 11 Audrey Snider-Bell/Shutterstock.com; p. 13 Evelyn D. Harrison/Shutterstock.com; p. 15 Agnieszka Bacal/Shutterstock.com; p. 16 Shcherbakov Ilya/Shutterstock.com; p. 17 Max Allen/Alamy Stock Photo; p. 18 Andriy Blokhin/Shutterstock.com; p. 19 Becky Sheridan/Shutterstock.com.

Printed in the United States of America

CPSIA compliance information: Batch #CSGS23: For further information, contact Gareth Stevens, New York, New York, at 1-800-542-2595.

CONTENTS

Words in the glossary appear in **bold** type the first time they are used in the text.

A UNIQUE ANIMAL

Australia is famous for its marsupials, such as the koala, kangaroo, and wombat. Marsupials are members of a **mammal** group in which the female carries her babies in a pouch on her belly. However, North and South America have their own marsupial: the opossum!

Opossums live mainly in the forest. However, they've **adapted** to living in cities and towns as well. It wouldn't be strange to see an opossum in your own backyard!

Australia has possums, which are very different from opossums.

AN OPOSSUM'S HABITAT

There are more than 60 species, or types, of opossum, and they all have different **habitats**. Most opossum species are found in South America and Central America. Some of these are small and look more like mice than like opossums in the United States.

The Virginia opossum is the only species whose habitat is north of Mexico. It's **native** to the eastern and central United States and as far north as southern Canada. It's not native to the western United States, but it lives there too!

FIND OUT MORE!

Scientists say all marsupials **evolved** in South America millions of years ago. At that time, South America and Australia were connected, so it was easy for the marsupials to get to Australia.

The Virginia opossum can be found in many parts of the United States. Its range is shown here in green.

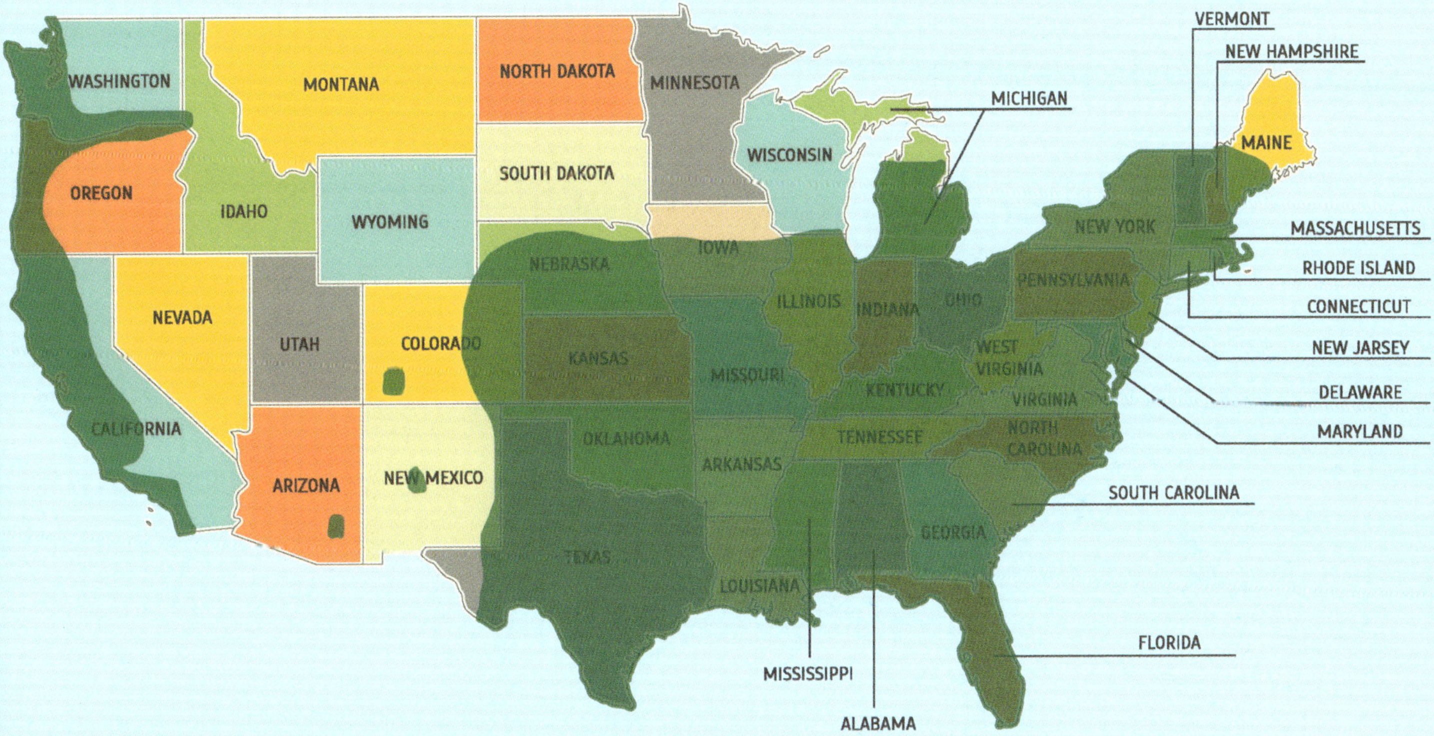

AT HOME IN THE TREES

Forests are great habitats for opossums because their bodies are built for climbing. Like some monkey species, opossums have prehensile tails. That means their tails can grasp and hold objects. Opossums can also grasp branches with their sharp claws. Their back feet have **opposable** thumbs, so they can also hold on with their feet.

Opossums are very adaptable and will make their den, or home, almost anywhere. This includes dens other animals have left behind, hollow tree stumps, and underneath buildings.

FIND OUT MORE!

Opossums have been around since the Upper Cretaceous period, when dinosaurs were still alive—and they haven't changed much in all that time!

Opossums can use their tails to hang upside down!

SPOTTING AN OPOSSUM

The Virginia opossum has a pointed snout, or nose and mouth, and a skinny tail. Its fur is generally dark gray on the body and white on the face. Female opossums are called jills. They have a pouch to carry their babies in, but males, or jacks, don't.

Jills and jacks are almost the same size. They're both about 16 inches (41 cm) long, not including their tail, which is about 11 inches (28 cm) long.

FIND OUT MORE!

An opossum has 50 teeth. That's more than any other mammal in North America! Opossums show those teeth when they start to feel **threatened**.

An opossum's front teeth are pointed and sharp.

JOEYS

Opossum mothers can have up to 25 babies at a time, but most have only about 8. These babies are called joeys. Opossum babies are only the size of a kidney bean when they're born. They're helpless without their mother.

After they're born, opossum babies have to crawl through their mother's fur and into her pouch. They stay there drinking their mother's milk for several weeks. After that, the young opossums climb in and out of the pouch.

FIND OUT MORE!

When the joeys are older, they ride around on their mom's back. At about three and a half months old, they leave their mother and start to explore on their own.

If a joey falls off its mom and doesn't get back to her, it goes off on its own.

A SPECIAL TRICK

Opossums don't like to fight. Instead, they have a special trick they can do. Opossums can make their body completely stiff, as if they're dead. They don't move when they're poked. Their bodies even start to smell dead.

Many of the opossum's predators, such as owls and red foxes, won't eat something that's already dead. If they think the opossum is dead, they'll often go away, and then the opossum is free to come back to life!

FIND OUT MORE!

Even though the proper name for this animal is opossum, many people simply call them possums. This is why "playing possum" is another way to say "playing dead."

An opossum can play dead for up to four hours!

NOT A PICKY EATER

Opossums will eat almost anything! They're omnivores, which means they eat both meat and plants. Some of the foods they like include fruits, nuts, bugs, and worms. They clean themselves like cats, and when they find ticks in their fur, they eat them. Sometimes they hunt small animals, such as mice or snakes.

Opossums are also scavengers, which means they'll eat dead plants and animals. Opossums also don't mind digging through people's trash to find food.

Opossums will eat what other animals leave behind.

LEAVE THEM BE

Because opossums sometimes drool and hiss when they feel threatened, many people think they have **rabies**. In fact, opossums almost never get rabies because their body temperature is too low for the virus to survive. They can carry other diseases, though.

If you see an opossum, it might show its many teeth and hiss at you to scare you away. However, opossums don't attack people unless they have no other choice. If you leave them alone, they'll go away.

Opossums only pretend to be **aggressive** so people and animals will leave them alone.

A HELPFUL ANIMAL

Opossums get a bad **reputation** because they're often found near tipped-over trash cans. However, they're usually just cleaning up the mess made by a larger animal, such as a dog or raccoon. They clean up messes in the forest too. By eating rotting animals, they stop diseases from spreading.

Opossums might not be as cute as some other animals, but they're very helpful to people. For example, they eat slugs, bugs, mice, and other **pests** that can ruin a garden.

FIND OUT MORE!

Eating ticks makes opossums very helpful to humans. Ticks can spread some nasty diseases to people. The fewer ticks there are, the less likely we are to get sick from them.

Opossums are really great to have around!

GLOSSARY

adapt: To change to suit conditions.

aggressive: Showing a readiness to attack.

evolve: To grow and change over time.

habitat: The natural place where an animal or plant lives.

mammal: A warm-blooded animal that has a backbone and hair, breathes air, and feeds milk to its young.

native: Living or growing naturally in a particular place.

opposable: Able to be placed against one or more of the fingers or toes on the same hand or foot, such as the human thumb.

pest: An animal or insect that causes problems for people.

rabies: A deadly disease that affects the central nervous system. It is carried in the spit of some animals.

reputation: The views that are held about something or someone.

threatened: Likely to be harmed.

FOR MORE INFORMATION

BOOKS

Rathburn, Betsy. *Opossums.* Minneapolis, MN: Bellwether Media, 2018.

Seed, Andy, and Claire Almon. *A Giant Dose of Gross: Discover the World's Most Disgusting Animals!* Lake Forest, CA: QEB Publishing, 2019.

WEBSITES

Easy Science for Kids: Opossums and Their Living Habits
easyscienceforkids.com/all-about-opossums
Read more about opossums and watch one play dead!

Kiddle: Opossum Facts for Kids
kids.kiddle.co/Opossum
Learn about opossums' diet, habitat, and more.

Publisher's note to educators and parents: Our editors have carefully reviewed these websites to ensure that they are suitable for students. Many websites change frequently, however, and we cannot guarantee that a site's future contents will continue to meet our high standards of quality and educational value. Be advised that students should be closely supervised whenever they access the internet.

INDEX